Navigating Performance Engineering

Volume 1:
The Collapse of Performance Value

First Edition

James L Pulley III

`https://www.perfbytes.com http://journeymanpublishing.com`

Navigating Performance Engineering- Volume 1: The Collapse of Performance Value. ©2026 by James Leonard Pulley, III

PerfBytes Press, An Imprint of Journeyman Publishing LLC

Errata: `Errors@press.perfbytes.com`

First Edition, 2026. Casebound: 978-1-964222-99-8
Paperback: 978-1-964222-07-3

Casebound editions are manufactured to institutional durability standards using library buckram and sewn signatures.

Names: Pulley, James Leonard, III, author.
Title: Navigating performance engineering.
Volume 1, The collapse of performance value / James Leonard Pulley III.
Description: Pauline, South Carolina : Journeyman Publishing LLC, 2026. | Includes bibliographical references and index.
Identifiers: LCCN 2026938311 ISBN 978-1-964222-99-8 (library binding) 978-1-964222-07-3 (perfect binding)
Subjects (LCSH): Software engineering—Risk management. Computer software—Reliability. Computer systems—Failure analysis. Measurement—Methodology. Systems engineering—Decision making. Organizational resilience.
Classification: LCC QA76.76.R5 P85 2026 DDC 005.1—dc23
LC record available at https://lccn.loc.gov/2026938311

P Journeyman Publishing

111 Foster Mill Circle, Pauline, South Carolina 29374
https://www.journeymanpublishing.com

*"The first principle is that you must not fool yourself —
and you are the easiest person to fool."*

— Richard Feynman

*"Most performance tests fail not because the system is
slow, but because the measurement was never worthy of
belief."*

— James L. Pulley

For the engineers who refused to turn measurement into theater, who understood that a test without integrity is worse than no test at all.

And for my wife, who endured more late nights and wandering thoughts than any book has a right to demand.

Contents

Acknowledgments

This volume would not have been possible without the feedback from a number of peers. I am forever grateful for the amount of time they spent reading earlier revisions of this work to provide feedback and insight.

Fuad Abu Safi

Tushar Baronia

Brian Brumfield

Carlos Chidiac

Sachin Dholakia

Nikolai Grabner

Robin Haynes

Yury Makedonov

Henry Steinhauer

Ádám Tóth

Preface

Y2K was the inflection point.

It was the moment when performance engineering stopped being an apprenticeship and became a staffing exercise.

We did not scale a discipline. We scaled headcount.

Seats were filled with people who could operate tools, not with people who could reason about systems. The objective quietly shifted—from understanding behavior to producing results, from engineering to execution, from truth to speed.

At the time, this felt necessary. The urgency was real. The risks were real. But something subtle and permanent was lost: the expectation that a performance professional must be able to move from a measurement to a decision, and from a decision to a change that improves a dimension of performance.

The industry began to confuse motion with progress.

Somewhere along the way, a dangerous sentence became acceptable:

> *"We don't have time for you to run a second test."*

That sentence is not about scheduling. It is about values.

It reveals a belief that speed matters more than certainty, that results matter more than integrity, and that a single measurement—no matter how fragile, no matter how unknown its repeatability—can be trusted if it arrives quickly enough.

It is the moment when performance work stopped being science and became theater.

There is a triangle that governs every engineering discipline:

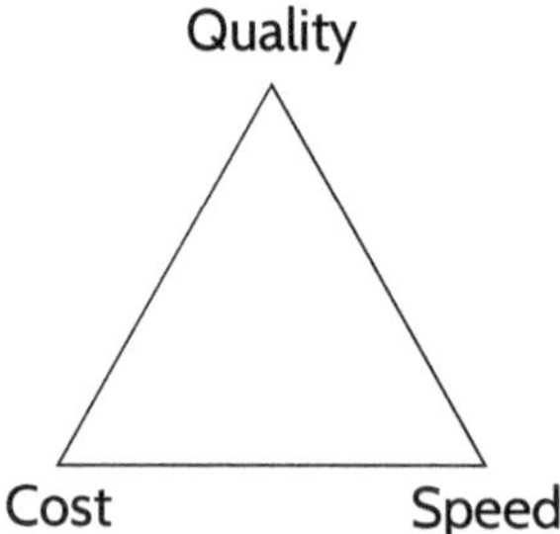

You may choose any two. You may never have all three.

The industry chose cost and speed and told itself it had preserved quality. Performance engineering became the quiet casualty of that compromise. Reproducibility was traded for schedule. Confidence was traded for convenience. Variance was ignored because it slowed the story.

And yet performance engineering was never a QA function.

It was always a business risk discipline with a quality attribute. Its purpose was not to confirm that a system worked. Its purpose was to reveal whether a system would fail under pressure—and how expensive that failure would be.

By recasting performance engineering as "just another test phase," we stripped it of its authority and then wondered why its warnings were ignored.

This book is written for three audiences at once.

It is written for performance engineers who continue to fight for measurement integrity even when their organizations reward speed over truth—those who remained curious when compliance would have been easier.

It is written for managers who sincerely want value, but who were never taught what value actually requires—who were handed teams and tools without a model for how performance work produces change, and who were later disappointed when nothing improved.

And it is written for tool vendors, whose products shape not only workflows but expectations. Tools do not merely enable behavior. They define what behavior appears possible.

This book is not an attack on tools. Tools are essential. Load generation, telemetry collection, and monitoring capabilities today are extraordinary technical achievements. We can generate more pressure, observe more systems, and collect more signals than at any point in history.

But execution power without measurement discipline does not produce value. It produces volume.

Test integrity is the foundation of performance engineering. Reproducibility is non-negotiable. Variance is not noise; it is the doorway to insight.

A test that cannot be trusted cannot inform a decision. A test that does not lead to change cannot deliver value.

In performance engineering, value is not speed. It is risk reduced before consequence arrives.

A performance practice that cannot translate measurement into defensible risk posture does not create value. It creates activity.

If nothing changes after a test, then the test has produced very expensive numbers.

This book is also written in defense of people who never chose this chaos. Many were socially promoted into performance roles, handed tools and applications without mentors, without training, and without a model for how value is actually delivered. They were told to "find performance problems" without being given the conditions required to discover them.

And yet they survived. They cared. They wanted to succeed.

The system failed them.

We are more dependent on software today than at any time in history. The fragility of our systems is increasing, not decreasing. The COVID pandemic showed the world what happens when scalability assumptions collapse under real stress. These failures were not academic. They were human. They affected healthcare, supply chains, government services, and basic trust in institutions.

And yet the profession responsible for anticipating those failures has been slowly hollowed out by its inability to demonstrate value under the conditions imposed upon it.

I wrote this book because I can no longer tolerate managers who do not understand how performance engineering creates value managing it in ways that make value impossible—and then expressing surprise when failures reach production.

This book is not a methodology. It is not a tool replacement. It is not an indictment.

It is a reconstruction.

It is an attempt to restore performance engineering as a discipline grounded in scientific thinking, economic consequence, and moral responsibility to truth.

Engineering disciplines survive through feedback loops. Observation produces signals. Orientation interprets those signals against models of reality. Decisions alter system structure. Actions generate new observations. When any stage of this loop collapses, disciplines drift from measurement toward instrumentation and from experiment toward assumption. Performance engineering did not fail because systems became complex. It failed because the feedback loop that maintained measurement integrity quietly broke.

If this book succeeds, you will not finish it with answers. You will finish it with better questions.

And that is exactly where engineering begins.

The Value Chain

How performance became a checkbox

1

Value, Defined

"Why wasn't it caught in performance testing?"

Performance engineering does not exist to produce reports. It does not exist to generate charts, or to proclaim that a system is "fast enough." Those are artifacts. Useful, sometimes impressive, often expensive. But they are not value.

Value is what happens *after* the measurement, when a system is changed. If nothing changes after a test, the test produced information, not value. If nothing changes after a quarter of testing, the organization has purchased very expensive numbers.

This book uses a strict definition of value, because without one the profession drifts into activity and away from impact. Performance work only becomes real when it intersects decision-making, implementation, and outcomes.

That strictness is not philosophical. It is protective. Vague definitions of value do not merely confuse reporting; they create environments where responsibility dissolves, where results cannot compel action, and where failure is always discovered too late to prevent harm. Precision is the only defense performance engineering has against becoming ceremonial.

A precise definition of value

Value, in this discipline, is the ability to recommend and implement a change in configuration, architecture, or code that produces a measurable improvement in a dimension of performance.

That definition is intentionally inconvenient. It requires interpretation, not execution. It requires ownership, not merely observation. It requires proof, not hope.

A dimension of performance is not limited to technical measures. Latency, throughput, stability, scalability, resiliency, and resource efficiency all matter— but they matter because they express themselves as business friction and business cost.

A slower checkout flow is not a problem because it is slower. It is a problem because it reduces conversion. An inefficient cache plan is not a problem because a header is wrong. It is a problem because it increases infrastructure cost and magnifies fragility during failure modes. A performance incident is not a problem because it produces ugly graphs. It is a problem because it creates support burden, operational distraction, revenue loss, and reputational damage.

Performance engineering lives at the intersection of technical behavior and economic consequence. Ultimately, business risk collapses to money. It either costs, or it makes.

Activity is not progress

Modern organizations are excellent at producing performance *activity*. They run tools, generate reports, hold readouts, and move tickets across boards. The presence of motion creates the feeling that something important is happening.

Activity becomes dangerous when it impersonates progress. Without an explicit connection to change, motion provides reassurance without control and visibility without authority.

The most common failure mode is not a bad tool, a slow script, or even a poor environment. It is the absence of a closed loop. The virtuous cycle of performance engineering is simple:

$$Measure \rightarrow Analyze \rightarrow Change \rightarrow Measure\ again$$

Remove the change step and the cycle collapses. Remove the final measurement step and confidence collapses.

When executives ask, "Why wasn't it caught?" they are not asking about tooling. They are asking why an organization paid for numbers that could not compel action.

When the Public Becomes the Monitor

There is one more truth that must be named before we proceed.

In the absence of disciplined pre-production measurement, organizations do not eliminate monitoring. They relocate it.

Today, the most expensive performance monitoring system deployed by many enterprises is not inside their infrastructure. It is the public.

When systems slow or fail, customers announce it in real time. They post screenshots. They compare notes. They amplify each other. Social platforms—particularly X—have become involuntary, distributed, real-user monitors.

This is not observability. It is consequence.

Unlike dashboards, public reaction does not wait for root cause analysis. It does not respect release schedules. It does not negotiate with executive narratives. It prices failure immediately in reputation, support burden, revenue impact, and market confidence.

When performance engineering fails to establish truth before release, truth is established afterward—by users, publicly, at scale.

The question is not whether monitoring exists. It always does.

The question is whether it exists inside a controlled experiment, or outside it.

Performance work that cannot survive before production will survive only as public spectacle.

And spectacle is the most expensive audit of all.

X is not a social platform in this context. It is a real-time, globally distributed outage reporting system that companies neither provisioned nor control.

When measurement fails privately, it succeeds publicly — at a price.

Performance engineering is not performance testing

Performance testing is often described as an extension of quality assurance. Performance engineering is not.

Quality assurance exists to confirm correctness. Performance engineering exists to reduce operational and economic risk under stress. One is concerned with whether a system works. The other is concerned with whether the system will fail, how it will fail, and what it will cost when it does. Performance engineering is, at its core, decision support for risk.

Tools are necessary in both worlds, but tools do not define the discipline. Engineers work with data. They interrogate variance. They demand repeatability. They treat measurement as evidence that must survive scrutiny.

For many organizations, the loss of faith in performance engineering did not arrive as a dramatic failure. It arrived slowly, as a pattern. Performance programs produced reports, dashboards, and executive summaries, yet the underlying systems continued to fail under real conditions. Releases were delayed without preventing outages. Budgets were spent without reducing operational uncertainty. Over time, leadership began to notice a troubling pattern: the discipline produced numbers, but those numbers rarely compelled action. When measurement cannot influence decisions, it stops being perceived as engineering and starts being perceived as ceremony. At that moment, the economic foundation of the discipline begins to erode.

What was quietly lost

Value cannot exist without truth. And truth, in engineering, is not a mood. It is something earned through method. The loss of rigor was not announced. It was experienced indirectly, as expectations shifted and shortcuts became normal.

Reproducibility was the first thread to fray. Once single executions were accepted as sufficient, the discipline stopped demanding that a result be repeatable before it was believed. A measurement that cannot be repeated is not evidence; it is anecdote.

When reproducibility weakens, control weakens with it. In any experiment, a control isolates causality. It allows "before" and "after" to mean something. It prevents drift from masquerading as discovery. Performance work slowly abandoned stable baselines and confirmation passes. Tests became events rather than experiments.

Then validation eroded. A successful HTTP status code began to substitute for correctness. "It returned 200" became synonymous with "it worked."

Systems can be fast and wrong, fast and incomplete, fast and degraded. Performance scripts that do not check expected outcomes measure speed without meaning.

FROM EXPERIMENT TO INSTRUMENTATION

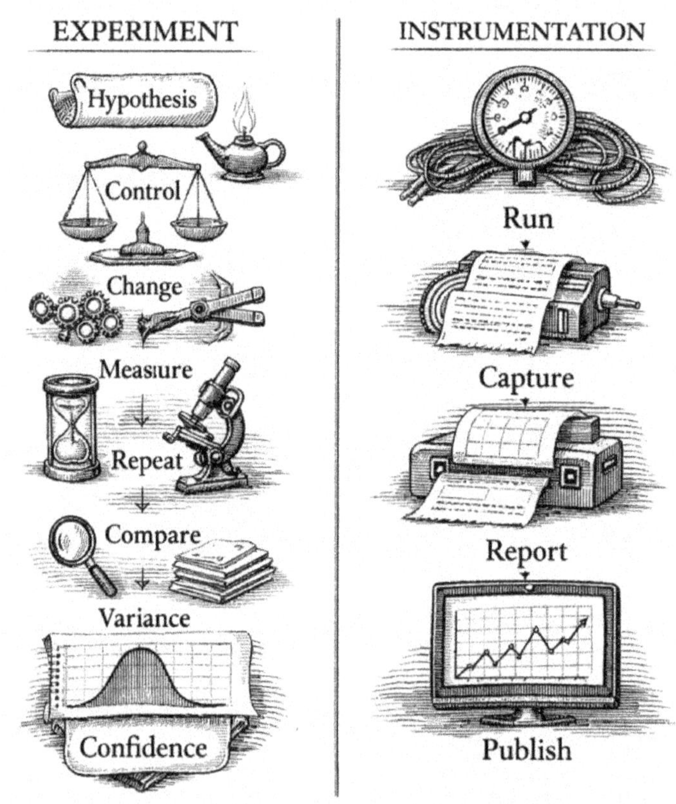

Performance engineering exists in the left column.
The industry increasingly operates in the right.

These losses—reproducibility, control, and validation—are not independent problems. They are the pillars of experimental thinking. Remove one, and the others weaken. Remove all three, and what remains is instrumentation, not science. Confidence becomes imagined, and risk decisions become guesses dressed up as metrics.

This did not happen because people chose to abandon rigor. In many organizations, it happened because people were never taught how rigor applies to performance work in the first place.

The profession expanded faster than its apprenticeship model could scale. Tool operation displaced experimental literacy. People did not violate the scientific method. They simply never inherited it.

A visual anchor: experiment versus instrumentation

The distinction matters because output can be produced quickly, cheaply, and repeatedly without ever becoming trustworthy. Truth is slower. It demands discipline. It demands that measurement survive repetition, control, and validation. It demands that we treat variance as information rather than noise.

The contract for this book

Throughout this book, value will mean only one thing: a change that makes a system better in a way we can prove. This definition is deliberately strict because it is the only one that protects the discipline from becoming a checkbox.

We will celebrate real technical achievements in tooling and observability. We will also be blunt about where the profession drifted away from scientific thinking, and why that drift made performance work easy to dismiss.

If the argument of this book can be reduced to a single sentence, it is this:

> *A test without integrity cannot guide change, and without change there is no value.*

What follows in the next chapter is not technical correction. It is an examination of how a discipline lost its inheritance, and what it cost the people asked to carry its responsibilities without its protections.

Performance engineering is not performance testing. Testing asks whether software works. Performance engineering asks whether software will fail under pressure and what that failure will cost.

It is measurement science applied to operational risk.

A Pause for the People

Before we continue, we need to pause.

Not for architecture. Not for tooling. Not for method.

For people.

We need to acknowledge the people who have lived inside the system that has been constructed around software performance engineering, and especially around performance testing.

Because what has been built over the last several decades is not just a technical system. It is a human system. And it has exacted a cost.

Performance engineers were asked to measure risk without being allowed to define risk. They were asked to validate quality without being given standards, to work scientifically without the protections of the scientific method, to move faster as cycle times compressed, even as the complexity of systems exploded, and asked to produce certainty in environments that were unstable, undocumented, and politically constrained.

And when results were inconclusive, inconsistent, or uncomfortable, those outcomes were not seen as evidence of a broken system. They were seen as evidence of inadequate execution.

This did not happen once. It happened repeatedly. Across organizations. Across projects. Across generations of practitioners.

Excellence did not fail. The system failed to recognize it.

What follows is not an argument about intent or effort. It is an examination of structure.

2

Talent, Mentorship, and the Inheritance Break

"With the right tool, any business analyst can be a performance
tester."

With the human cost acknowledged, we can now examine the structural failure that made it inevitable.

Before Y2K, performance engineering was not something you staffed. It was something you inherited.

Every organization that ran serious systems long enough eventually had one of them. Not a title. Not a role on an org chart.

A function.

Every organization had a Glenn—the curmudgeon in the corner who knew where the code bodies were buried, and why.

Glenn was not introduced as "the performance engineer." Glenn was introduced as "the one who knows the system."

He remembered why a configuration had been changed during a crisis and never revisited. He knew which database table grew faster than anyone expected. He understood that a sudden increase in memory pressure was not a resource problem, but an architectural confession.

This mattered, because performance engineering was never about execution alone. It was about memory.

Performance tools in that era were built for Glenn. Not to replace judgment, but to amplify it. They accelerated hypothesis testing, shortened feedback loops, and made intuition survivable by anchoring it in data.

The understanding lived in the person. The tool merely extended its reach.

This was an apprenticeship model. It transferred judgment, not procedure. It preserved skepticism, not compliance. It taught scientific habits implicitly, because the work could not succeed without them.

Reproducibility was not optional because Glenn would not accept a result he could not repeat. Controls mattered because Glenn demanded to know what had actually changed. Validation mattered because Glenn refused to accept speed without correctness. Variance mattered because stability was treated as a signal, not an inconvenience.

This model was slow. It was expensive. And it worked.

Then Y2K happened.

Demand for performance professionals exploded overnight. Every organization was suddenly forced to validate behavior at scale. Every timeline compressed. Every system was placed under suspicion.

The apprenticeship model could not scale at that speed.

There was no time to grow Glenns. No time to transfer lineage. Only time to put bodies in seats.

So the definition of "qualified" changed.

Before: *Can reason about systems.*

After: *Can operate the tool.*

That shift appears subtle. It is not.

It marks the moment when performance engineering moved from judgment-based authority to tool-based legitimacy.

This was not a philosophical decision. It was an operational necessity.

But operational necessities are never neutral. They reshape disciplines whether those disciplines are prepared or not.

Vendor messaging accelerated the shift. Demonstrations showed complete performance tests built in hours. Record. Correlate. Execute. Report.

The implication was unmistakable: the hard parts had been solved. Judgment had been embedded. Expertise had been automated.

What actually happened was subtler—and more damaging.

This is where social promotion entered the profession.

Capable people from adjacent roles were promoted into performance responsibilities. They were given accountability without inheritance. Responsibility without authority. Risk without insulation.

They were handed tools and systems and told to deliver certainty.

This was not mentorship. It was risk transfer.

Enterprise risk was moved from organizations onto individuals without changing governance, mandate, or protection. The failure mode was built in.

When outcomes were favorable, the tool received the credit. When outcomes were inconclusive or uncomfortable, the individual absorbed the blame.

The vendor narrative had already prepared the ground: "If anyone can do this, failure must be personal."

Most practitioners sensed the gap immediately. They felt the difference between execution and understanding. They knew they were accountable for decisions they were not equipped to defend.

But the system rewarded output, not skepticism. Speed, not rigor. Compliance, not truth.

So they adapted.

They learned to execute. They learned to report. They learned to survive.

What they were never taught was how to ask whether the measurement itself was true.

This was the first fracture in authority. When the profession could no longer defend the integrity of its own measurements, it ceased to be advisory and became procedural.

This is how a discipline becomes performative.

Not through incompetence. Through structural inevitability.

When a discipline loses its elders, it does not merely lose people. It loses memory.

When judgment cannot be inherited, it is replaced with procedure. When memory disappears, tools are asked to remember. When science becomes inconvenient, instrumentation impersonates it.

Some practitioners resisted.

They insisted on retests. They demanded controls. They refused fragile results.

They did not save the discipline. They masked its collapse.

As elders left, memory went with them. And when a discipline loses its memory, tools stop extending judgment and begin replacing it.

That inversion is where the next failure begins.

FROM APPRENTICESHIP TO ASSIGNMENT

<table>
<tr><th>APPRENTICESHIP
(PRE-Y2K)</th><th>ASSIGNMENT
(POST-Y2K)</th></tr>
</table>

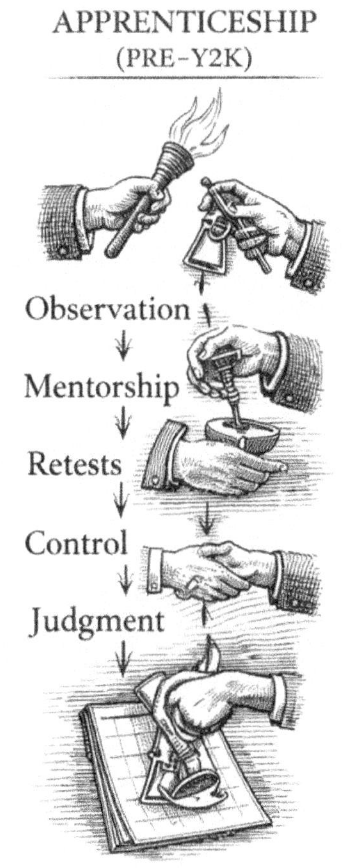

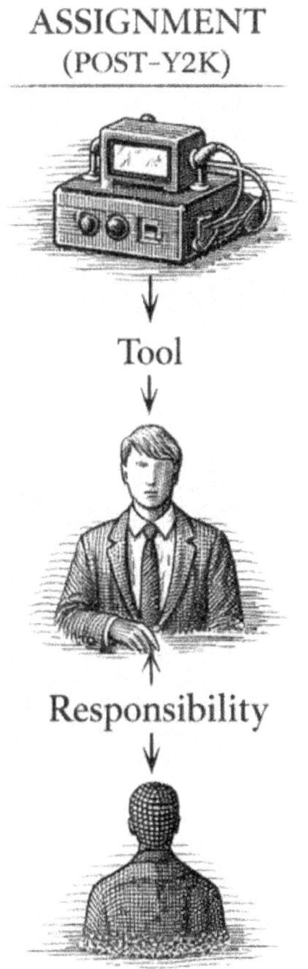

One creates engineers. The other creates operators.

Apprenticeship transfers judgment. Assignment transfers liability.

3

When Tools Replaced Measurement

Tools were not always judges. There was a time when performance tools were instruments—extensions of human reasoning, not substitutes for it. They existed to make measurement easier, faster, and more defensible, but they never defined what measurement meant. They assumed that the person using them already understood systems, statistics, and failure.

Tools were microscopes, not judges.

They helped confirm hypotheses. They helped visualize patterns. They helped turn intuition into evidence.

They did not claim authority.

That authority belonged to the engineer.

The inversion began quietly, and it began after the inheritance break.

When judgment could no longer be reliably inherited, it was replaced with procedure. When institutional memory thinned, tools were asked to remember. Measurement stopped being something constructed and became something received. The profession did not notice when this happened because the output still looked impressive. Charts were produced. Dashboards were filled. Reports were published. The machinery of performance continued to run even as its purpose quietly changed.

At first, the shift seemed harmless. Faster execution was welcomed. Broader access felt democratic. The idea that more people could participate in performance testing appeared to be progress.

But access without education does not create empowerment. It creates dependency.

This is the moment when instruments became oracles.

Before, a performance result was a starting point for interrogation. After, a performance result became an answer.

The difference is subtle and devastating.

When an engineer uses a tool, the question is:

What does this tell me about the system?

When a tool replaces the engineer, the question becomes:

What does the tool say?

Those two questions are not equivalent.

One treats measurement as evidence. The other treats measurement as verdict.

This shift was reinforced by vendor messaging that emphasized ease of use and speed of execution. The promise was irresistible: sophisticated performance testing without sophisticated performance thinking. Demos showed entire test scenarios built in an hour.

Scripts recorded, correlations applied, load executed, reports generated. The implication was unavoidable: the hard parts had been solved. Expertise had been automated.

But what was automated was not understanding. What was automated was output.

The ability to generate traffic does not imply the ability to interpret behavior. The ability to generate a report does not imply the ability to evaluate risk. The ability to execute a test does not imply the ability to change a system.

Tools did not replace engineers. They replaced skepticism.

This inversion was compounded by the collapse of mentorship described in the previous chapter. When people were promoted into performance roles without training, they inherited responsibility without inheritance. They were given powerful tools but no scientific foundation for using them. The tool became their teacher. And tools teach only what they encode.

What was not encoded was reproducibility, control, variance, confidence, validation, and causality. So those concepts disappeared from practice.

Monitoring suffered the same fate.

Before the shift, monitoring was investigative. It was performed by people who understood systems and expected instability. They looked for deviation. They compared behavior across time. They distrusted averages. They asked why a system behaved differently under load, under time, under configuration change.

Monitoring was where performance engineering and system science met.

Once tools were treated as judges, monitoring could no longer function as investigation. Dashboards replaced inquiry. Green lights replaced reasoning. "Looks fine" replaced "what changed?" The goal became reassurance, not understanding. Monitoring became a display of stability rather than a search for instability.

Variance and the cost of discomfort

Statistics is not everyone's favorite subject. Many people carry scars from introductory courses that emphasized formulas over meaning. For some, the phrase "standard deviation" evokes anxiety more quickly than curiosity.

That discomfort matters, because variance is not optional in performance engineering. Variance is where behavior reveals itself. Averages are summaries. Variance is evidence of instability, contention, drift, and unknown change. When a discipline quietly agrees not to look where it feels least comfortable, it does not become simpler—it becomes blind.

Untrained people did not monitor because they did not know what monitoring was for. Untrained people did not use standard deviation because no one had taught them that averages lie. Untrained people did not look for variance because variance only matters when you understand behavior.

So when developer-centric tools emerged, they inherited a world already stripped of statistical literacy. They were built in an environment where averages were considered sufficient, single runs were accepted, pass/fail was preferred to confidence, and speed mattered more than truth.

The features that slowed execution disappeared: confirmation runs, baseline validation, distribution analysis, and statistical stability. Not because they were wrong, but because they were inconvenient.

This is how tools stopped serving measurement and started defining it.

A performance test became something you run. Not something you construct.

A result became something you receive. Not something you earn.

A dashboard became something you trust. Not something you challenge.

This was not the fault of the tools. It was the fault of the discipline for allowing its thinking to be externalized.

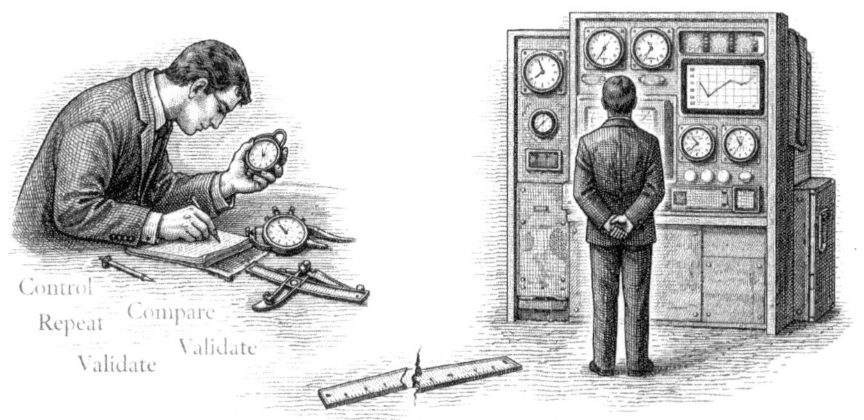

When tools stopped serving measurement and started defining it.

When tools become oracles, organizations stop asking whether results are true. They ask whether results exist. The mere presence of data becomes validation. The mere presence of a chart becomes confidence. The mere presence of green becomes safety.

This is why performance failures now feel mysterious. This is why incidents prompt disbelief. This is why the question returns again and again:

"Why wasn't this caught?"

Because tools were never designed to catch truth. They were designed to present output.

Truth requires human skepticism. Truth requires variance. Truth requires the willingness to distrust convenience.

When a discipline forgets this, it does not merely lose its tools. It loses its authority.

4

Management, Cycle Time, and the Compression Trap

Performance engineering did not fail because it was deprioritized. It failed because it was positioned at the end of the process.

And anything placed at the end of a delivery pipeline inherits the failures of everything before it.

Functional testing slipped. Integration slipped. Environments slipped. Data preparation slipped.

Performance did not slip. It was compressed.

By the time performance testing was scheduled, time was no longer a resource. It was a liability. The question was never, "How much time does this require to be done well?" It was always, "How quickly can we get through it?"

Just get it done. Check the box. Is it fast enough?

This is how a scientific discipline becomes a ceremony.

Performance work was not planned. It was absorbed. It became the shock absorber for schedule failure. And shock absorbers are not designed for accuracy. They are designed for impact.

When functional testing ran long, performance testing became shorter. When environments were late, performance testing became smaller. When data was incomplete, performance testing became shallower.

No one sat down and decided that quality should be reduced. The system simply taught everyone that quality was negotiable and time was not.

That is how cycle time quietly became the dominant virtue.

Managers were rewarded for shipping. Teams were rewarded for closing tickets. Projects were rewarded for hitting dates.

No one was rewarded for asking whether the results were true.

Here is the hidden hinge: many managers were never taught where performance value is actually created. They were shown tools. They were shown dashboards. They were shown elapsed time. What they were not shown was the labor required to produce a defensible change—the cycles of measurement, analysis, revision, and confirmation that transform numbers into decisions.

Performance value does not map cleanly to labor hours. It arrives after false starts, discarded hypotheses, repeated runs, and uncomfortable conclusions. That shape is difficult to forecast and difficult to govern with a plan. So management retreated to what was legible: task duration and labor allocation. Not malicious. Not deliberate. A proxy chosen in the absence of a clear value model.

The performance phase became a gate, not an investigation. A step to clear, not a problem to solve. And once performance became a gate, its purpose inverted. It no longer existed to reduce risk. It existed to authorize progress.

This is where the phrase "fast enough" was born.

"Fast enough" does not mean correct. "Fast enough" does not mean safe. "Fast enough" does not mean stable.

It means: *we can move on.*

And moving on was what the system valued.

Performance testing is code. It has a planning window. It has dependencies. It has complexity. Scripts are programs. Load models are specifications. Data is a dependency. Environments are infrastructure. Monitoring is instrumentation. Analysis is engineering.

But performance work was rarely governed as engineering. It was scheduled as execution.

"Can you build the scripts in the morning and test in the afternoon?"

That question only makes sense if accuracy is optional and results are cosmetic. It treats engineering as button-pressing and science as configuration.

This was not stupidity. It was misclassification. Performance engineering was treated as execution when it was, in truth, investigation.

Managers were rarely shown that performance engineering obeys the same laws as any other form of engineering: design time, construction time, validation time, and revision time. Without those phases, what remains is not engineering. It is theater.

Once time disappeared, reproducibility disappeared with it.

You cannot repeat what you do not have time to run again. You cannot control what you cannot isolate. You cannot confirm what you cannot revisit.

The first test became the only test. The first result became the truth. Variance became noise instead of signal.

This is where confidence collapsed.

Without repeatability, you cannot know if behavior is stable. Without stability, you cannot know if change was real. Without change, you cannot claim value.

So performance results became symbolic. They existed to satisfy process, not to guide action. The work was no longer about discovering weakness. It was about demonstrating progress.

And in that environment, management behavior was predictable.

Managers optimized for what they could see: schedule adherence, ticket completion, green dashboards, tool output.

They were rarely shown: confidence intervals, distribution stability, variance collapse, risk amplification.

Quality was abstract. Risk was invisible. Speed was measurable.

What is measurable wins budget. What is defensible wins authority. Performance engineering lost both.

So speed won.

This is the classic triangle: cost, speed, quality. You can choose any two. You never get all three.

Performance engineering lives in the space where quality must be protected even when speed is rewarded. But once performance was moved to the end of the pipeline, that protection vanished.

There was no buffer left. There was no margin for science.

"We don't have time to run it again."

That sentence is the death of measurement integrity.

Not because repetition is bureaucratic, but because repetition is how truth is established. Without repetition, a result is not evidence. It is a guess wearing numbers.

This is how performance testing became a checkbox. Not because people stopped caring. But because the system stopped making room for caring to matter.

By the time performance ran, the decision to ship had already been made. The test existed to validate momentum, not to challenge it. Failure was unacceptable because failure meant delay, and delay meant cost.

So results became curated. Scope became minimized. Assumptions became protected.

The profession did not collapse under ignorance. It collapsed under urgency.

Urgency is not automatically negligence. But urgency left unexamined produces the same outcome: decisions made without truth, and risk carried without consent.

Not malicious negligence. Structural negligence.

Compressed timelines rarely feel like choices. They feel like facts. And when they are treated as facts, uncertainty becomes unwelcome.

This is why managers often feel betrayed by performance failures. They were told the system had passed. They were shown green. They were shown speed. What they were not shown was uncertainty.

The Unknown Unknowns of Governance

The people defining performance scope did not know what they did not know. The people funding performance work did not know what they were not being shown. When governance operates without an explicit model of measurement integrity, ignorance compounds silently.

Performance engineering cannot exist where uncertainty is unwelcome.

Because uncertainty is where risk lives.

5

Requirements Without Truth

Performance testing without requirements is not testing. It is observation.

This distinction sounds semantic until its consequences are felt. When no operational performance requirement exists, every measurement becomes a matter of opinion. Each team carries a different internal definition of "fast enough." Architecture imagines one threshold. Development imagines another. Operations imagines a third. Quality assurance imagines a fourth. Performance engineering is left attempting to reconcile competing beliefs rather than evaluating objective truth.

In this environment, performance testing becomes an exploratory ritual: *"Run the test and see what it does." "Tell us if it looks slow."*

But slow relative to what? Fast compared to which standard? Acceptable under what operational contract?

Without requirements, performance has no anchor. It floats.

This is why production incidents are so often followed by the same question:

Why wasn't this caught in performance testing?

The answer is rarely technical. It is almost always epistemological. There was nothing to measure against.

This is not merely a performance problem. It is an operational requirements problem.

Performance, accessibility, and security share a common trait: they do not emerge naturally from feature discussion. They must be asked for explicitly, or they do not appear at all.

Functional requirements are straightforward. If a user performs action X, outcome Y is expected. The behavior is visible. The correctness is concrete.

Operational requirements are different. They govern experience under load, failure, and time. They describe constraints, not features. And constraints are easy to overlook when no one is accountable for naming them.

In modern development environments—particularly those that emphasize iterative feature evolution—conversation naturally gravitates toward what can be seen. User interfaces dominate discussion because they are tangible. Performance rarely does, because its absence is silent until it becomes catastrophic.

This is not a castigation of agile development. It is a recognition that questions not asked do not produce requirements that govern behavior.

It is also tempting to treat this as a problem introduced by modern methods. That would be incorrect.

Earlier lifecycle models were not consistently effective at capturing operational requirements either. They produced specifications, but those specifications were often silent on performance, accessibility, and security beyond vague statements of intent. Waterfall created the appearance of rigor through documentation. Agile removed that appearance in favor of adaptability. Neither model, by default, repaired what was missing.

The industry changed how software was built without correcting a long-standing omission: the failure to make operational truth explicit, testable, and owned.

When performance is not named as a requirement, it cannot be tested as truth. It can only be observed after the fact.

Test Versus Benchmark

A test and a benchmark are not the same thing, though the industry frequently treats them as interchangeable.

A **test** exists to validate a hypothesis:

> The system will meet a defined performance requirement under defined conditions.

A test requires:

- A requirement

- A hypothesis

- Controlled execution

- Repeatability

- Interpretation against expected outcomes

A **benchmark** exists to observe:

> This is how this configuration behaves under a standardized workload.

Benchmarks produce measurements. Tests produce judgments.

Benchmarks answer:

Are we faster or slower than before?

Tests answer:

Did we meet the requirement?

Without requirements, benchmarks are mistaken for tests. Trend lines are mistaken for truth.

What TPC and SPEC Got Right

The Transaction Processing Performance Council (TPC) and the Standard Performance Evaluation Corporation (SPEC) represent some of the highest discipline in performance measurement ever created.

Their documentation is not merely procedural. It is philosophical.

Take TPC-C in particular. It defines:

- Data models

- Transaction mixes

- Measurement rules

- Reporting constraints

- Auditing requirements

- Disclosure obligations

TPC-C is not "run the test and see." It is a contract with truth.

And yet, TPC benchmarks are still benchmarks. They are not production requirements. They are standardized workloads intended to create comparable measurements across platforms.

This distinction matters deeply.

When organizations treat benchmarks as surrogates for requirements, they replace operational truth with competitive theater.

The Ugly Baby Problem

When a benchmark produces an uncomfortable result, a predictable reaction follows:

> The benchmark must be wrong.

This is the ugly baby problem.

In the absence of a shared, explicit performance requirement, every group arrives with its own internal understanding of load, scale, and acceptable behavior. Architects design with one mental model. Developers implement with another. Operations plans for a third. Each model is internally consistent. None of them are written down. None of them are reconciled.

When a performance test is executed under assumptions that differ from those internal models, the result is not merely surprising. It is perceived as incorrect.

The test is not evaluated as evidence. It is evaluated as an accusation.

This response is not irrational. It is human.

Code is not abstract. It is authored. It represents decisions, tradeoffs, and effort. When a test highlights a weakness, it does not feel like a neutral measurement. It feels personal. The result challenges not just the system, but the assumptions that guided its construction.

So the instinctive response is defensive:

> That load isn't realistic. Users don't behave that way. The environment is wrong. The test is flawed.

And in many cases, those objections are partially true. The assumptions used to build the code may differ from the assumptions used to conduct the test. But without an agreed-upon requirement, there is no external standard to resolve the conflict.

At that point, disagreement is inevitable.

Like a parent confronted with an unflattering photograph, teams rush to the scene: *My baby is not ugly. You're wrong. You need glasses.*

This is not deception. It is not incompetence. It is what happens when judgment is attempted without a contract.

Benchmarks become proxies for requirements. Tests become proxies for truth. And any result that conflicts with prior belief is treated as suspect.

This is why requirements matter. They externalize expectation. They remove interpretation. They allow disagreement to be resolved by reference, not by authority or emotion.

Without that anchor, performance testing does not reveal truth. It reveals ownership.

Why Requirements Change Everything

A performance requirement is a promise:

> This system will deliver this experience under these conditions.

Requirements unify:

- Architecture design

- Code implementation

- Infrastructure sizing

- Caching strategy

- Data model design

- Capacity planning

Without them, each group optimizes against an imagined target. With them, the organization shares a single operational truth.

When a test fails, there is no debate. The system failed to meet its contract.

When a test passes, the organization gains confidence not because the numbers look good, but because the numbers have meaning.

The Cost of Ambiguity

Without requirements:

- Performance becomes subjective

- Accountability dissolves

- Teams defend rather than improve

- Data loses authority

- Testing becomes theater

The profession does not fail because engineers lack skill. It fails because truth is never defined.

Measurement and Engineering

At this point in the chapter, the reader must internalize a final distinction:

> Observation is not engineering. Comparison against a standard is engineering.

Measurement without a reference is observation.
Measurement against a standard is engineering.

The Boundary

This chapter defines the boundary between:

- Reporting and engineering

- Data and truth

- Numbers and accountability

Benchmarks observe. Tests judge.

And judgment is impossible without requirements.

In God we trust. All others must bring data.
W. Edwards Deming (attributed)

6

Environments and Data

Every performance environment is a declaration.

It declares what an organization believes about truth. It declares what it is willing to fund. It declares whether measurement is expected to be scientific or merely symbolic.

For decades, performance engineers were handed environments that looked like laboratories and were told to act like scientists inside them. They were asked to draw conclusions, to assess risk, to provide confidence. But the environments themselves were not designed to reveal truth. They were designed to be convenient.

This was not a tooling failure. It was a moral one.

By moral, this does not mean malicious. It means structural. It means choosing reassurance over falsifiability.

A laboratory is not defined by its purpose. It is defined by its honesty.

In science, honesty means this: a system must be capable of being proven wrong.

A real laboratory preserves causality. Change one variable and observe one outcome. Hold everything else stable.

Most performance environments did the opposite.

They altered many variables at once and then asked engineers to explain results as if causality still existed.

Shared infrastructure. Reduced hardware. Mocked dependencies. Missing tiers. Different storage classes. Different network topology. Different caching layers. Different data.

These were not "smaller versions of production." They were different systems.

And different systems do not answer the same questions.

The Lie of Equivalence

The lie was subtle.

"We're just scaling it down."

But scale is not linear.

You can only scale load to your most limiting component. If your database is at one-tenth production capacity, your entire environment is at one-tenth capacity no matter how large everything else is. If your application tier is smaller, the whole system is smaller. If your storage is slower, the whole system is slower.

The smallest element defines the ceiling.

That alone would be survivable if environments were uniformly scaled. A uniformly smaller system still preserves ratios. It still exhibits proportional behavior. It still allows science.

But environments were rarely uniform.

Non-uniformity is worse than uniformity.

When systems are scaled unevenly, emergent behavior disappears. Queuing effects vanish. Contention masks itself. Collapse points relocate or never appear.

The system no longer behaves like a system.

When you squeeze the foot of a balloon animal, nothing changes in the rest of the balloon. The pressure does not redistribute. The shape lies.

That is exactly what happens when:

- The database is production-grade but the application tier is tiny

- The application tier is production-grade but storage is mocked

- The CDN is removed or bypassed

- Authentication services are stubbed

- Third-party dependencies are bypassed

You are no longer observing a system. You are observing isolated components that cannot interact naturally.

Causality collapses.

At that point, performance testing becomes theater. Not because people intend deception, but because physics is no longer present to enforce honesty.

Data Completes the Illusion

Data completes the illusion.

Small datasets behave politely. They fit in memory. They eliminate disk I/O. They collapse lock contention. They remove index pressure. They hide fragmentation. They erase cache churn.

They also lie about time.

Small datasets do not age. They do not fragment. They do not accumulate historical scars. They do not reveal long-tail behavior or slow decay.

A system with small data is not a system. It is a demonstration.

And performance engineers were asked to certify risk based on demonstrations.

They were handed environments that could never reveal failure modes and then blamed when those failure modes appeared in production.

This is why trust eroded.

Not because engineers were incompetent. But because they were asked to prove truth in laboratories that could not produce it.

A Historic Shift: On-Demand Truth

But something fundamental has changed.

For the first time in the history of computing, exact environments are economically possible.

With cloud infrastructure, we can:

- Replicate production topology

- Replicate production scale

- Replicate production storage classes

- Replicate production network behavior

- Replicate production cache layers

- Replicate production data volumes

for short, controlled windows.

This is unprecedented.

The historical excuses were once valid:

- Hardware was scarce

- Provisioning was slow

- Cost was prohibitive

- Risk was high

Those constraints are gone.

Capability, however, is not intent. The ability to know does not guarantee the willingness to look.

The era of fake laboratories is over. The only remaining question is whether organizations are willing to pay for truth before they pay for failure. We just have not fully updated our thinking.

Microscope and Wind Tunnel

Full replication is not the only scientific tool available.

Just as single-user truth reveals defects invisible under concurrency, single-node stack architectures reveal behavior invisible at scale.

One node of each architectural tier:

- One web node

- One application node

- One database node

- One cache node

With a single user and a network tap, you can observe:

- Chatty interfaces

- Over-allocation

- Early serialization

- Protocol inefficiencies

- Redundant calls

- Unnecessary payload sizes

This is not a compromise. It is a microscope.

Cloud-scale environments then become the wind tunnel:

- They expose contention

- They expose collapse points

- They expose nonlinear failure

- They expose queuing behavior

Together, they form a real laboratory: precision and scale. Microscope and wind tunnel.

For decades, performance engineers were denied both.

They were given stages and asked to perform science on them.

The Verdict

This chapter is not about infrastructure. It is about honesty.

An environment is a moral object. It encodes whether an organization wants truth or comfort.

If you fund a real environment, you are saying:

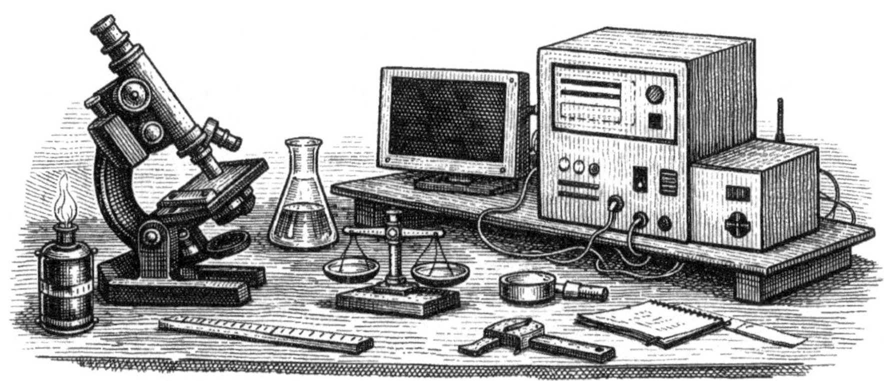

A false laboratory produces false confidence.

We want to know, even if it hurts.

If you fund a toy environment, you are saying:

We want reassurance, not reality.

And performance engineers lived inside that contradiction for a generation.

Not because they lacked skill. But because they were never given a laboratory.

7

Execution Integrity and Clocks

There is a moment in serious performance work when the numbers do not fail—but trust does.

Not loudly. Not catastrophically. Not in a way that immediately signals error.

Measurements continue to arrive. Dashboards still render. Percentiles still compute. Confidence intervals still exist.

And yet something fundamental has shifted.

The same test is run again. The workload is held constant. Nothing that should matter is changed. And the results move anyway.

Not by tenths of a percent. By whole percentages. Sometimes more.

At first, this is treated as noise. The Internet is involved. Distributed systems are variable by nature. Some movement is expected.

But reproducibility is not a preference. It is the foundation of science.

If results cannot be reproduced within a narrow tolerance band, then the experiment is no longer interrogating a system. It is sampling instability.

What makes this failure dangerous is that it does not look like failure.

It looks like variance.

For decades, performance engineering trusted time without thinking about it.

That trust was inherited.

Physical clocks were owned. They were monotonic. They were local. They were boring.

They did not drift meaningfully. They did not pause. They did not negotiate with schedulers. They did not borrow time from somewhere else.

Because time had never betrayed the discipline, it was never governed. No requirement was written for it. No control was established. No validation step checked whether time itself was behaving honestly.

Time was assumed to be part of the background—like gravity.

That assumption held for a very long time.

Then virtualization arrived.

Quietly. Incrementally. Productively.

And it broke an invariant the discipline never knew it was relying on.

Virtual machines do not own time. They borrow it.

A virtual machine's clock is not a physical instrument. It is a negotiated construct, shaped by:

- Host scheduling

- CPU steal time

- Pause events

- Clock drift

- Resynchronization events

When a virtual machine falls behind, the hypervisor intervenes. When it jumps forward, the measurement absorbs the distortion silently.

Nothing crashes. Nothing alerts. Nothing looks wrong.

But the timeline has warped.

When a timing record is open and the clock bends:

- That transaction appears longer

- Percentiles skew upward

- Averages inflate

- Standard deviation expands

- Confidence collapses

Not because the application slowed. Not because the system degraded.

Because time stopped being trustworthy.

At that moment, performance engineering is no longer measuring performance. It is measuring clock instability.

Virtual Machine Clock, in Dalivision

This knowledge has existed for years.

It lives in hypervisor documentation, kernel discussions, and specialist forums. It is discussed carefully, quietly, and precisely.

What it has not been translated into is operational consequence.

Organizations are not told:

Your measurements may be invalid.

They are told:

VM clocks can drift under certain conditions.

That sounds academic. Optional. Inconsequential.

It is none of those things.

Time is not an implementation detail. It is the substrate of every performance metric in use:

- Response time

- Throughput

- Variance

- Percentiles

- Confidence intervals

- Saturation curves

They are all functions of time.

Break time, and measurement collapses.

When measurement collapses, confidence becomes interpretive. When confidence becomes interpretive, decisions become political. If your clocks are unstable, your confidence intervals are fiction.

This chapter is not an argument against virtualization. Virtualization is not the failure.

The failure is assuming that time remained invariant when the execution substrate changed.

Execution integrity means the act of measurement itself must be trustworthy.

Not the scripts. Not the load. Not the system under test.

The clock.

If time cannot be trusted, nothing built on top of it can be trusted.

This is why control elements matter.

A physical load generator becomes a reference. A known-stable clock becomes a baseline. A fixed execution substrate allows variability elsewhere to be interpreted.

Without controls, every test becomes self-referential. You cannot distinguish system change from measurement distortion.

This extends beyond clocks.

Virtualized monitoring introduces similar integrity failures:

- Negative CPU usage

- Impossible memory patterns

- Bursts that violate physical constraints

- Metrics that reflect host contention rather than system behavior

Dashboards remain precise. They lie elegantly.

This is not negligence. It is unexamined inheritance.

Execution integrity demands that time be treated as sacred.

Just as chemistry requires calibrated scales, and physics requires stable clocks, performance engineering requires a temporal substrate that does not lie.

When organizations say:

It shouldn't matter,

what they are really saying is:

We do not believe time is part of the experiment.

But time *is* the experiment.

When time becomes elastic, results become interpretive. When results become interpretive, confidence becomes political. When confidence becomes political, performance engineering ceases to be engineering.

Execution integrity begins with trustworthy time.

Because when time melts, truth melts with it.

And no amount of tooling can save you.

If time cannot be independently validated, audit collapses before it begins.

8

Observability Without Navigation

*Not everything that can be counted counts, and not everything
that counts can be counted. — William Bruce Cameron*

There are moments in engineering where progress is so profound that it quietly redefines what is possible. Observability has been one of those moments.

For the first time in the history of large-scale computing, we can see software systems not as static diagrams or theoretical flows, but as living organisms. We can watch requests traverse networks; we can observe contention emerge and dissolve; we can trace causality across service boundaries; we can measure user frustration in milliseconds and pixels.

Real-user monitoring exposed the lived experience of customers with unprecedented clarity. Deep diagnostics revealed what was previously hidden behind kernels, schedulers, and runtime abstractions. eBPF opened a window into operating system behavior that had been inaccessible without invasive instrumentation. Distributed tracing stitched together cause and effect across hundreds of services.

These were not incremental improvements. They were foundational.

They changed software engineering from inference to observation, from storytelling to evidence, from assumption to measurement.

But this chapter is not a celebration.

It is an explanation of why observability became economically inevitable.

As confidence in pre-production performance risk management eroded, organizations did not abandon measurement. They relocated it.

They relocated it to the one place where truth could not be debated: production.

In pre-production, measurement is always vulnerable to the same objections:

- The environment is not equivalent.

- The data is not real.

- The clocks are not stable.

- The workload is not representative.

- The requirements were never explicit.

Those objections are not always cynical. Sometimes they are correct.

But when an organization cannot reliably produce defensible truth before release, it will produce truth after release.

That truth is expensive. It arrives as:

- customer frustration,

- support burden,

- operational disruption,

- and reputational damage.

Observability grew because it could convert those consequences into evidence.

Production observability is not merely visibility. It is authority.

And markets will always fund authority.

Then came OpenTelemetry.

Before OTEL, observability lived inside fractured dialects. Each vendor spoke its own language. Instrumentation was proprietary. Switching tools meant rewriting code. Semantics diverged. Data models conflicted. Engineering teams became locked into telemetry ecosystems.

OTEL unified that chaos.

It provided a shared grammar, a shared transport, and a shared structure. It did not replace vendors; it reduced the cost of switching. It did not suppress innovation; it standardized communication.

The telemetry Tower of Babel collapsed into a common dialect. Imperfect, perhaps, but usable at planetary scale. An Esperanto for engineering truth.

This did not make observability cheap. It made observability transferable.

And transferability is what markets demand when dependence becomes costly.

Navigation Does Not Scale

Observability scales infrastructure. Navigation scales only with people.

This is the tension most organizations do not name.

Collecting telemetry can be automated. Interpreting it cannot.

The ability to generate traces does not create the ability to reason about them. The ability to store logs does not create the ability to form hypotheses. The ability to render dashboards does not create the ability to identify causality.

Navigation is skilled human judgment applied to evidence. It is the capacity to move from *data* to *decision*. It is scarce. It does not improve linearly with more tools. It scales with apprenticeship, practice, and time.

When organizations treat navigation as if it scales like telemetry, they do not just create inefficiency. They create risk.

Observability provides stars. Navigation provides meaning.

When the Observer Becomes Heavy

For a brief moment, instruments and practice aligned.

Then success scaled.

Telemetry became easier to collect. Storage became cheaper. Pipelines became faster. Instrumentation became standardized. And observability stopped being merely a window into systems and became a system of its own.

Entire platforms emerged to manage telemetry pipelines. Entire teams formed to tune retention policies. Entire budgets became dominated by the cost of watching other budgets operate.

In some environments, the volume of observability data now exceeds the volume of customer data. The observer has become heavier than the observed.

With mass comes gravity.

Telemetry consumes CPU, memory, network bandwidth, disk I/O, and engineering attention. Observability is no longer just a technical decision. It is an economic decision.

This is where distortion begins.

Sampling is no longer chosen because it preserves scientific rigor. It is chosen because it preserves budgets.

Retention is no longer chosen for forensic completeness. It is chosen for financial survival.

Truth becomes shaped by what is affordable to store, query, and retain.

Observation Perturbs the System

As observability industrialized, collection became automated. Interpretation did not. Navigation remained human.

The navigator is now asked not only to interpret complexity, but to do so under constraints imposed by the act of observation itself.

We have entered the era of *quantum monitoring*.

In physics, observation perturbs the system being observed. In performance engineering, instrumentation perturbs latency and resource utilization. In modern observability, telemetry perturbs economics, behavior, and decision-making.

Observation now changes:

- the technical system,

- the operational system,

- the financial system,

- and the risk system.

We are no longer merely measuring reality. We are shaping which realities are affordable to see.

This is not a tooling problem. It is a governance problem.

Because when observation becomes expensive, truth becomes selective.

The Boundary

Observability remains one of the great triumphs of modern engineering. OTEL remains one of its most elegant achievements. But neither observability nor OTEL can substitute for navigation.

Telemetry scales. Judgment does not.

When organizations fund instruments but do not fund navigators, they do not eliminate risk. They relocate it.

And relocated risk always accrues interest.

When observation becomes expensive, truth becomes selective.

And when truth becomes selective, confidence becomes political.

9

The Missing Audit

The first principle is that you must not fool yourself — and you
are the easiest person to fool. — Richard P. Feynman

We have talked about tools. We have talked about talent and mentorship. We have talked about environments, data, clocks, observability, and navigation.

Now we have to talk about something far more uncomfortable:

How do we know whether any of this work is *good*?

Not emotionally good. Not politically acceptable. Not impressive on a slide deck.

Scientifically sound.

Because without an audit model, performance testing is not engineering. It is theater.

This loss of auditability did not occur because engineers became careless or indifferent. It occurred because no organization made survivability of truth an explicit requirement.

The question beneath the question

There is a sentence every organization eventually speaks, usually after production behaves differently than the report predicted:

"Why wasn't this caught in performance testing?"

That question is not really about speed, scale, or even correctness. It is about confidence. It is the moment when an organization realizes it never established what kind of evidence it was willing to accept.

It is the sound of two standards colliding: the standard of storytelling and the standard of science. One relies on trust. The other relies on reproducibility.

Every time someone asks, *"Why wasn't this caught?"* what they are really asking is: *"Why did we believe something that could not survive independent verification?"*

The gate: can it survive the loss of the author?

Here is the definition that matters:

If another qualified group cannot reproduce your work and your results without trusting you, then what you produced was not an experiment. It was a demonstration.

And more bluntly:

If your performance test cannot survive a taco truck losing its brakes on a steep hill, it is not science. It is a story that happened to work while you were alive to defend it.

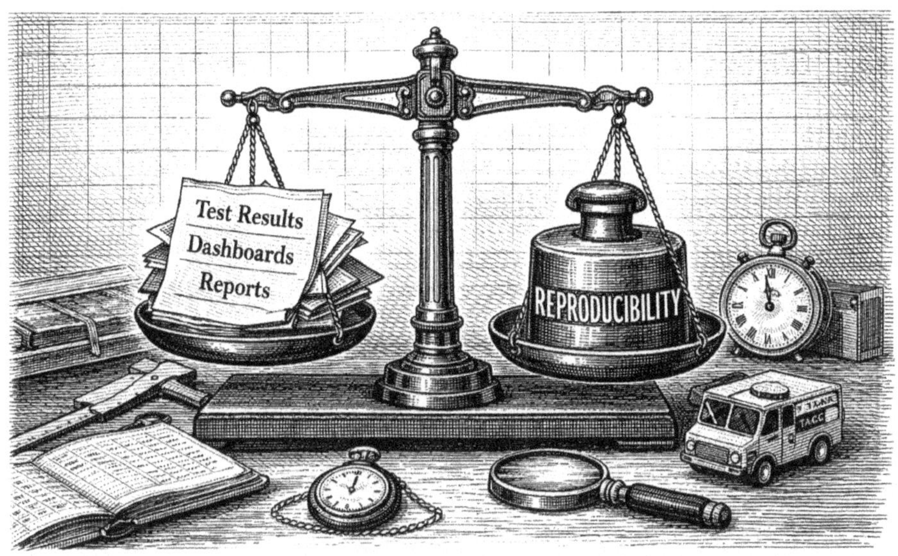

Truth must outweigh trust.

That line is not humor. It is a boundary condition.

Science must survive accidents. Engineering must survive turnover. If your results die with you, they were never results. They were narration.

Reproducibility is not a maturity level. It is a gate. If your work cannot survive independent reproduction, it should not be allowed to influence risk decisions.

Why TPC and SPEC exist

This is why benchmarks like TPC and SPEC exist. Not primarily to measure performance, but to defend truth from human nature.

Their designers understood something uncomfortable:

Engineers will lie to themselves.

Not maliciously. Not consciously. But inevitably.

We simplify models. We reuse convenient data. We accept partial environments. We rationalize missing controls. We adjust workloads "just this once". We explain away variance.

Under schedule pressure, budget pressure, or career pressure, we begin optimizing the *result*, not the experiment.

TPC and SPEC are built on this assumption. They do not rely on personal integrity. They rely on structure.

They require rigor so that a published result can be defended in two directions:

1. Against the environment itself — proving the system behaved that way under defined conditions.

2. Against independent third parties — proving the result was not shaped through convenient omissions, biased assumptions, or a chain of individually reasonable but compounding decisions.

What internal performance testing kept, and what it lost

Internal performance testing kept the tools and lost the audit cage.

We kept:

- load generators

- scripts

- dashboards

- reports

We lost:

- environment equivalence

- clock verification

- dataset reconstruction rules

- statistical tolerance enforcement

- peer review

- independent reproduction

So performance testing quietly shifted from engineering to belief:

"Trust me, I ran it."

But trust is fragile. Trust collapses under pressure. Trust evaporates when production behaves differently.

Science does not ask for trust. Science asks for reconstruction.

The minimal audit model

A minimal audit model is not documentation theater. It is the ability for strangers to rebuild what you did without relying on your presence.

Can it repeat?

If the system is reset and the workload executed again, does the result fall within a defensible range?

If not, something in the environment is variant:

- Capacity

- Network path

- Time base

- Data state

- Concurrency profile

- Hidden dependency

Non-repeatability is not inconvenience. It is signal.

Until a measurement repeats within known bounds, it cannot be aligned to business risk.

- Environment: Can they reconstruct the hardware, software, and configuration?

- Data: Can they regenerate scale and distribution?

- Workload: Can they reproduce the load model and pacing?

- Artifacts: Can they recreate scripts and collectors?

- Monitoring: Can they reproduce monitors and sampling?

- Measurement: Are clocks stable and timing rules explicit?

- Statistics: Is variance measured, and are tolerance bands declared?

If any of these are artisanal, undocumented, or dependent on personal memory, reproducibility collapses.

And when reproducibility collapses, credibility collapses with it.

Audit does not slow you down; it makes speed honest

Auditability does not reduce speed. It removes illusion.

Without auditability:

- speed produces fragility

- efficiency produces blind spots

- tooling becomes a shield

- results become persuasion

With auditability:

- confidence is earned

- disagreement becomes productive

- failure becomes diagnosable

- performance engineering becomes durable

The missing foundation

This is the discipline's missing foundation:

Results must survive the loss of their authors. Results must survive skepticism. Results must survive independent reconstruction.

Until performance engineering accepts that its output must be reconstructable without personal authority, it will remain dependent on individual excellence compensating for systemic absence of auditability.

And fragile truth is not truth at all.

10

Risk Without Measurement Science

Risk is not a number. It is not a response time. It is not a throughput curve. It is not CPU utilization, memory pressure, or error rate.

Risk is the distance between what we believe and what is actually true, multiplied by how much damage that gap can cause.

When organizations talk about "performance risk", they often mean "performance unknowns". The implicit belief is that if enough data is gathered, risk will eventually reveal itself. But data without measurement science does not reduce risk. It manufactures confidence. And confidence without structure is how failures become catastrophic.

Before examining the examples that follow, one assumption must be softened: that failure implies negligence or incompetence. In performance engineering, failure is more often the product of absence.

Absence of mentorship. Absence of exemplars. Absence of any lived experience of what performance engineering looks like when it consistently delivers value.

Most organizations have never seen it. Not once. Not in a decade. Sometimes not in an entire corporate lifetime.

What they have seen instead are performance teams that arrived late, produced charts, spoke in percentiles, generated dashboards, and left without changing a single architectural decision. They have seen testing that produced numbers but not outcomes. They have seen risk reported but never reduced.

Over time, those experiences form a baseline. Performance becomes ritual. Noise. A cost center that produces discomfort but not clarity.

So when real performance engineering appears, it is not recognized as expertise. It is filtered through disappointment. That is not malice. It is conditioning.

This conditioning underlies every failure that follows.

Risk grows in the gap between belief & truth

Healthcare.gov: the risk of absence

The Healthcare.gov launch remains one of the most public examples of performance failure in modern software. On its first day, only six users successfully completed enrollment. Political shock followed, emergency engineering teams were assembled, and the site became a symbol of technical dysfunction.

But the failure was not architectural. It was epistemological.

The system was designed around a core human assumption: that users would convert on their first visit. In reality, most people require multiple visits before entering sensitive personal information. That single assumption multiplied load expectations by orders of magnitude.

The risk was not hidden. It was never framed as a hypothesis to be tested.

Six weeks before launch, I asked a peer network of mature performance architects whether anyone was working on the project. None were involved. None were even aware of who was.

That silence was information.

The absence of mature performance engineering talent is itself a measurable risk signal. Healthcare.gov did not fail because tools were missing. It failed because measurement science was absent.

Coinbase: the illusion of modernity

In February 2022, Coinbase aired a Super Bowl advertisement: a floating QR code drifting across a black screen. No branding. No copy. Just curiosity.

Scan it.

Every scan sent traffic to the same landing page. One URL. One entry point. One synchronized surge of traffic. It was marketing brilliance and performance catastrophe waiting to happen.

This was not a scaling failure. It was a failure of load geometry. A failure to recognize that human synchronization is an amplifier. Millions of independent users behave identically when given the same stimulus at the same time.

One question would have changed the outcome:

"What happens if everyone arrives at once?"

A developer test does not answer this. A benchmark does not answer this. Past success does not answer this.

Only measurement science does.

Coinbase had observability. It had dashboards. It had metrics. What it lacked was anticipation. Observability cannot prevent what engineering never imagines.

Salesforce: when measurement threatens authority

In 2017, I joined a company using Salesforce deployed in Europe as its global instance. Data protection concerns drove the decision. Performance was assured to be a non-issue.

It was an issue everywhere outside Europe.

A multigenerational redirect chain across the Atlantic amplified latency by orders of magnitude. The physics were obvious. The solution was obvious. But the original decision-maker was in the room.

Measurement threatened narrative. Performance engineering became a challenge to authority, not a technical discussion.

The findings were minimized. Recommendations rewritten. Failures misattributed. Individuals discredited. Not because the engineering was wrong, but because being right carried political cost.

Years later, the instance was relocated. Performance improved dramatically. The engineering had always been correct. It simply lacked survivability in a room where authority mattered more than accuracy.

What these stories share

Healthcare.gov shows the risk of absence. Coinbase shows the illusion of modernity. Salesforce shows the political gravity that suppresses measurement.

All three share the same root cause:

They collected data without measurement science. They trusted numbers without understanding variance. They relied on confidence without confidence bounds. They measured activity instead of risk.

Performance engineering did not fail to protect them. It was never given the authority to do so.

The mandate

Risk is not reduced by numbers. It is reduced by knowing how wrong the numbers might be.

Measurement science is what gives performance engineering authority:

- reproducibility

- variance modeling

- confidence intervals

- scenario sensitivity

- human behavior modeling

- independence from individual heroics

Without it, performance work becomes theater. Charts replace truth. Dashboards replace accountability. Confidence replaces evidence.

These stories are not indictments of people. They are indictments of absence.

Until performance engineering becomes measurement science, it will remain powerless in rooms where power matters more than accuracy.

11

The Cost of Late Discovery

Time is the most expensive bug multiplier. — Field axiom

Every engineering discipline eventually converges on the same truth: defects are not equally expensive at every point in time. They obey an economic curve. The earlier a flaw is discovered, the cheaper it is to correct. The later it is discovered, the more violently its cost escalates.

This is not a software principle. It is a property of reality.

Medicine understands it. Manufacturing understands it. Aviation understands it. Structural engineering understands it.

Software, for a long time, pretended it was exempt.

We told ourselves that code was malleable, that rollback was easy, that patches were cheap, that the cloud would forgive us. We behaved as if time did not multiply cost. We behaved as if performance failures could be absorbed at the end of the pipeline without consequence.

What we built instead was a system where truth is delayed until the moment when it is most expensive to hear.

Performance technical debt accumulates until the first performance question is asked. If that question is asked in production, the organization has selected the most expensive location to discover the most expensive class of defect.

Not because production is morally wrong. But because production is where cost is maximized.

In production, architecture is frozen. Data gravity exists. Customers are present. Revenue is flowing. Reputation is fragile. Recovery is public. Every defect has an audience.

A problem discovered here is no longer an engineering issue. It is a business event.

This is why late discovery feels catastrophic. It is not that the defect is worse. It is that the context is unforgiving.

A single-request failure, discovered at the worst possible moment

Consider J.Crew, Black Friday 2018.

Marketing placed a new image on the front page. It was visually striking. High resolution. Rich color. Perfect composition.

It was over forty megabytes in size. It was not optimized. Not compressed. Not cacheable. It did not live in the CDN.

The result was immediate. The site slowed and then collapsed on the highest traffic day of the year during the most valuable revenue window.

This was not a concurrency failure. It was not a database scaling failure. It was not a missed load test.

It was a single-request engineering failure that discovered itself in production.

The fix was trivial:

- Resize the image

- Reduce color depth

- Optimize the format

- Make the asset cacheable

- Serve it from the CDN

Minutes of work. Discovered at the worst possible moment.

The cost was not only lost transactions. Performance failures do not merely lose revenue. They change behavior. They reshape trust. They redirect loyalty.

Performance technical debt does not just accumulate. It compounds socially.

Surprise is the invoice

FinOps literature often frames cost as a function of usage. But its deepest truth is simpler:

Cost is not created by usage. Cost is created by surprise.

Performance failures are surprise generators. They convert uncertainty into invoices. A system that cannot measure performance risk early does not avoid cost. It postpones payment to the moment when interest is highest.

Performance technical debt behaves like financial debt:

- It compounds quietly

- It hides while growing

- It becomes catastrophic when collected

Production is where the collection agency operates.

This is why canaries are not prevention. They are damage detectors. They tell you the bill has arrived. They do not tell you the debt was avoided.

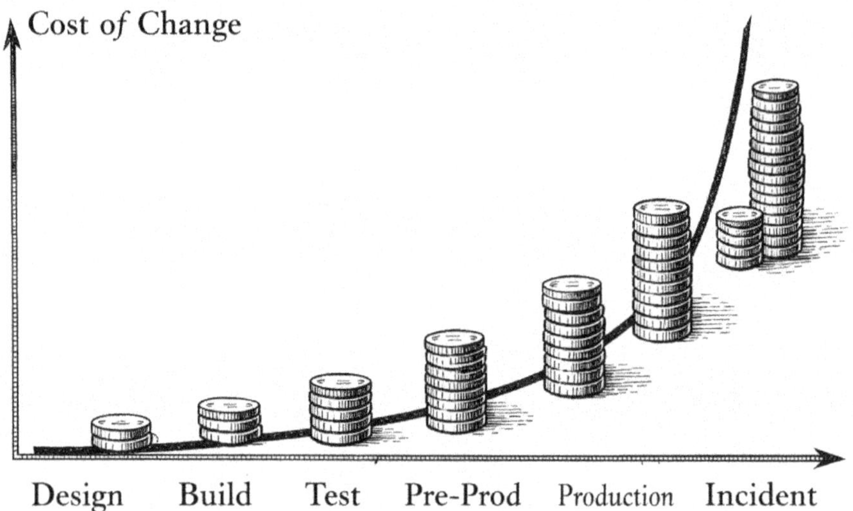

Time is the most expensive bug multiplier

Tools shape where truth is discovered

If late discovery is economically catastrophic, then the systems we use to discover truth are not neutral. Tools shape economic exposure. They influence where organizations believe truth should be found.

Most performance tools are historians. They reconstruct failure beautifully. They answer:

"What broke, and how badly?"

Almost none are designed to answer the only question a business actually needs answered:

"What failure will cost us the most, and how do we prevent it before it exists?"

This is not a criticism. It is lineage. Performance tools evolved from debugging, not economics. They were born when performance was a technical problem, not a financial exposure.

The world has changed.

To the tool vendors reading this

The question is not adversarial. It is an invitation:

What would your product look like if it were designed to prevent economic loss, not to visualize technical failure?

A risk-first tool would organize reality by revenue sensitivity, blast radius, irreversibility, time-to-fix, and trust erosion.

Dashboards show damage. Risk engines prevent it.

Microscopes explain failures. Telescopes prevent disasters.

The close

Performance testing is not expensive. Performance ignorance is.

Every organization will pay for performance truth. The only question is when, and at what interest rate.

Every tool that moves discovery earlier is not just a technical product. It is a financial instrument.

12

What the Market Got Right — and the Gap That Remains

The chapters before this one have been demanding. They have asked the reader to sit with uncomfortable truths about process erosion, scientific drift, misplaced incentives, and the slow substitution of dashboards for understanding. This chapter exists for a different reason.

It is here to acknowledge that the modern performance and observability ecosystem is, in many ways, a remarkable achievement. What exists today would have been unthinkable even a decade ago. We did not stumble into this capability. It was engineered. It was built by people who cared deeply about visibility, reliability, and operational truth.

The market did not fail. It succeeded—brilliantly—at solving a different problem than the one we are now asking it to solve.

That success was not accidental. It was economic.

As early-stage performance engineering failed to consistently deliver confidence before production, value migrated. Organizations did not stop caring about truth. They relocated where they expected to find it. Production observability became the place where uncertainty could finally be resolved—because it was the only place left where answers reliably appeared.

The value had to go somewhere. Observability was the natural destination.

The Observatory

If performance engineering were astronomy, then the industry has built a world-class observatory.

We now have sensors capable of capturing phenomena once considered invisible. We can trace requests across hundreds of services. We can measure latencies at nanosecond resolution. We can observe memory pressure, CPU contention, cache eviction, kernel scheduling, network congestion, and application-level state transitions in real time.

OpenTelemetry unified a fragmented universe of telemetry into a common language. What once required bespoke collectors, incompatible formats, and heroic integration work now flows through standardized pipelines. The tower of Babel was replaced by Esperanto, and for the first time, everyone could speak to one another in data.

This was not an incremental win. It was a civilizational one for the discipline.

Observability today gives us *stars*. It shows us vast, beautiful, complex systems in exquisite detail. It gives us the raw materials of truth.

But an observatory without navigation is still just a sky full of light.

We solved Stability. Navigation comes next.

The observatory flourished not because navigation was solved, but because navigation had failed elsewhere. When early risk discovery became unreliable, organizations invested where truth still emerged. Production visibility delivered answers—even if those answers arrived late and at high cost.

The Race Already Won

The industry has mastered:

- Data capture

- Data transport

- Data storage

- Data visualization

- Data correlation

Those are not small accomplishments. They represent decades of cumulative engineering labor. They are victories earned under real constraints.

Yet notice what is absent from that list:

- Measurement integrity

- Experimental control

- Repeatability

- Variance interpretation

- Risk translation

- Decision accountability

The market got the instruments right. It did so because instruments delivered immediate, visible value.

What remains unfinished is not a skills gap. It is an incentive gap.

Measurement science delivers its value by preventing loss, not by producing spectacle. Until failure occurs—or nearly occurs—its contribution is invisible. Markets reliably optimize for what can be demonstrated, not for what quietly removes danger.

The Folded Flag

Every engineering race has a finish line. Every discipline needs a moment where work transforms into meaning.

The folded victory flag is not about celebration. It is about completion.

It represents:

- A system sufficiently understood

- Risks that have been consciously accepted or mitigated

- Decisions that can be explained and defended

- Outcomes that survive scrutiny, time, and consequence

Observability shows us motion. Measurement science tells us when motion becomes progress.

Victory is not the absence of defects. Victory is knowing which risks were taken—and why.

The Gap

The gap is not technological. It is epistemological.

We know how to gather data. We do not yet know how to universally transform it into decisions that survive pressure, time, and accountability.

We have built telescopes. We have not finished writing astronomy.

This book does not seek to diminish what has been built. It seeks to finish it.

The future discipline of performance engineering will not be defined by how much we can see.

It will be defined by how confidently—and defensibly—we can decide.

13

Laying the Track

We have spent this volume naming things that most professions prefer not to name.

We named how value became rare. We named how speed displaced integrity. We named how measurement drifted into observation. We named how tools became substitutes for thinking. We named how reporting replaced accountability. We named how confidence was assumed instead of earned.

And in doing so, we did something more important than criticism: we restored clarity.

Because clarity is the beginning of health.

This book has not been an indictment of people. It has been an indictment of structure.

We have shown that performance engineering did not fail because it lacked intelligence, effort, or intent. It failed because it was asked to operate as a scientific discipline without being given scientific infrastructure. Engineers were asked to defend truth without being given systems that make truth defensible. They were asked to manage business risk without being given mechanisms to measure risk scientifically. They were asked to deliver confidence using tools designed to produce numbers, not certainty.

That is not a human failure. It is an architectural failure.

You cannot ask individuals to carry the weight of a discipline that has not been structurally built.

We ask performance engineers to behave like scientists while denying them reproducibility, control, auditability, variance modeling, confidence thresholds, and economic translation. We ask them to deliver risk insight while surrounding them with systems that reward speed, convenience, and visual comfort instead of truth. We then judge them for failing to produce outcomes that the system itself makes impossible.

This volume has shown how that happened. It has shown how well-intentioned tools, processes, and organizational pressures slowly displaced scientific rigor with mechanical execution. It has shown how partial successes were mistaken for completion. It has shown how an entire profession became instrument-rich and structure-poor.

And yet, embedded in this diagnosis is something profoundly hopeful:

Nothing here is unsalvageable. Nothing here is irreparable. Nothing here requires heroism.

It only requires engineering.

Every mature discipline becomes healthy the same way: not through motivation, not through training alone, not through better dashboards, but through infrastructure.

Railroads did not become reliable because engineers worked harder. They became reliable because track standards were enforced. Because load limits were known. Because signaling was formalized. Because failure modes were understood. Because logistics was engineered.

Medicine did not become scientific because doctors were more moral. It became scientific because protocols were standardized. Because controls were introduced. Because reproducibility was demanded. Because evidence became portable between institutions.

Performance engineering is standing at that same threshold.

What it lacks is not vision. What it lacks is a spine.

A spine is not a metaphor for strength. It is a metaphor for logistics.

A spine carries load. A spine aligns movement. A spine protects integrity. A spine enables growth without collapse.

Without a spine, power becomes chaos. Without a spine, speed becomes danger. Without a spine, effort becomes noise.

The measurement spine is the logistics backbone that performance engineering has never had.

It is the system that determines: what constitutes valid data, when data is trustworthy, how variance is interpreted, when confidence is sufficient to act, how measurements become economic risk signals, how results can be audited, how truth survives organizational change, how value is delivered without heroics.

This is not about replacing tools. It is about governing them.

This is not about adding work. It is about removing ambiguity.

This is not about changing people. It is about changing systems so people no longer have to compensate for missing structure.

And this is where value begins to return to its rightful owners.

Users regain trust when systems behave predictably under stress. Vendors regain credibility when results are defensible, not performative. VARs regain authority when guidance is grounded in science, not tooling allegiance.

Ownership re-emerges when confidence is earned, not asserted.

Steel, rock, and wood do not symbolize speed. They symbolize permanence.

Railroad track is not glamorous. It is heavy. It is slow to build. It is unforgiving when laid incorrectly. And once placed, it determines everything that follows.

That is what you are about to build.

Volume II does not argue that performance engineering *should* be scientific. It assumes that it must be.

It does not ask whether confidence matters. It defines how confidence is constructed.

It does not suggest that risk can be measured. It shows how risk becomes measurable.

It does not ask people to behave better. It builds systems that make integrity automatic.

Volume I was diagnosis. Volume II is construction.

This is where we stop talking about what went wrong and start engineering what must exist.

This is where performance engineering stops being a collection of tools and becomes a discipline with logistics.

This is where truth gains track.

Performance engineering is not testing. It is measurement science applied to operational risk. Volume 1 has shown how that science was gradually replaced by instrumentation, dashboards, and convenience. Diagnosis is not the end of a discipline. It is the beginning of its repair. Volume 2 begins the restoration of the feedback loop that keeps engineering honest.

Afterword

This work was never meant to be a conclusion. It was meant to be a recognition.

Recognition of the profession we inherited. Recognition of the systems we operate inside. Recognition of the quiet frustration many of us have carried without a language to describe it.

If you have read this far, it is not because you are curious about performance testing as a topic. It is because somewhere in these pages, you recognized yourself. In the tension between speed and integrity. In the pressure to deliver answers without being given the conditions necessary to make those answers trustworthy. In the moments where you knew something was wrong, but could not always prove it in a way that would be heard.

This book does not place blame. It names patterns. It does not accuse. It explains. It does not diminish the work that has been done. It honors how difficult that work has always been.

We built a profession under extraordinary pressure. We built it while tooling evolved faster than education. While management expectations compressed faster than process maturity. While the demand for certainty grew even as the foundations for certainty weakened. The fact that meaningful work was accomplished at all under these conditions is not a failure of the profession. It is evidence of its resilience.

And yet resilience alone is not enough.

We have lived inside a system where speed was rewarded more consistently than truth, where delivery was measured more often than value, where dashboards were celebrated more readily than insight. Many of us adapted in order to survive. Some of us fought to hold the line. Some of us quietly left for other disciplines where rigor was easier to defend.

None of those responses were wrong. They were human.

This book exists because something deeper was missing. Not more tools. Not more dashboards. Not more automation. What was missing was a shared agreement that performance engineering is not an act of measurement alone, but an act of responsibility. That it is not about producing numbers, but about protecting decisions. That it is not a QA function, but a risk function grounded in science.

We are not trying to return to an older era. We are trying to move forward with clarity.

The future of this profession does not come from rejecting modern observability, automation, or scale. It comes from giving those capabilities a backbone. From anchoring them to reproducibility. To variance. To control elements. To environments and data that can be trusted. To requirements that mean something. To economics that can be defended. To human systems that are supported rather than exhausted.

That future is not built by any one person. It is built collectively—through shared language, shared expectations, and shared courage to say when something is not yet ready to be called science.

If this book has done anything, it has created a space where that conversation can be honest.

We are not behind. We are early in defining what "good" actually looks like.

We are not broken. We are under-specified.

We are not failing. We are learning how to measure what matters.

Volume II will not begin something new. It will give shape to something that already exists inside this community: the instinct to protect integrity, the desire to deliver value, the refusal to accept fragile truth as sufficient.

The work does not begin in Volume II.

It began the moment you recognized yourself in these pages.

On Lineage

This book is written in the language of modern systems: tools, pipelines, environments, telemetry, and clocks.

But the discipline it defends is much older than software.

The argument in these pages is not new. It is not a reaction to cloud computing, or to Agile methods, or to the economics of observability platforms. Those are merely the current stage on which an ancient conflict is being reenacted:

Do we want reassurance, or do we want truth?

Long before anyone measured latency, people argued about evidence. Long before anyone built dashboards, people argued about interpretation. Long before anyone ran tests, people argued about how we know what we claim to know.

The Greeks did not give us "the scientific method" as a branded artifact. They gave us something more fundamental: the insistence that claims must survive questioning.

Socrates did not leave behind a checklist. He left behind a posture: that confident statements should be interrogated until they either become stronger or collapse. That posture is uncomfortable. It is also the beginning of integrity.

Plato gave language to the distinction between appearance and reality. Software produces endless appearance: green dashboards, fast averages, successful status codes, polished narratives. The work of performance engineering begins when appearance is treated as insufficient.

Aristotle insisted on causality. Not merely that something happened, but why it happened; not merely correlation, but structure. Modern systems tempt us to accept stories because there are too many moving parts to isolate the truth. Causality is the antidote to that temptation. Control elements, baselines, and repeatability are simply Aristotle with instrumentation.

Hippocrates is often remembered for medicine, but what matters here is the moral framing: a profession is accountable for the harm it prevents, not merely the activity it performs. Performance engineering is not morally neutral when it claims confidence. A false assurance is not a neutral outcome. It is an engineered risk.

Archimedes reminds us that measurement is not a vibe. It is geometry, definition, and constraint. A lever works because its properties are stable. A scale works because its reference is stable. A timing record works only if time itself can be trusted.

This lineage matters for one reason:

It makes clear that the demands of this book are not modern preferences. They are ancient requirements for defensible knowledge.

Reproducibility is not bureaucracy. It is the minimum condition for claiming that a result is real.

Variance is not nuisance. It is information about a system's stability and the measurement's integrity.

Auditability is not paperwork. It is the requirement that truth survive the loss of the narrator.

Performance engineering does not need to become something new. It needs to remember what it already is when it is honest.

The tools will change. The architectures will change. The delivery models will change. The vendors will change. The dashboards will change.

But the underlying standard does not change:

A claim that cannot survive questioning is not knowledge. A result that cannot survive reconstruction is not science. A measurement that cannot survive time, control, and validation is not truth.

This book is not asking the reader to adopt an ideology. It is asking the reader to rejoin a lineage.

And that lineage is old.

Very old.

Old enough that it does not care what your tool reports. Old enough that it does not accept "good enough" as an argument. Old enough that it will still be here when our current systems are replaced.

That is the kind of stability a profession needs when it intends to matter.

Colophon

The book is composed in a 7×10-inch format with extended outer margins to support annotation, slow reading, and long-form technical study. All illustrations are rendered in an engraved, scientific pen-and-ink style to reflect the visual language of early engineering and scientific texts.

This work leverages a classical bookmaking style in which all chapters begin on a right-facing page. This will occasionally produce a blank facing page on the left. As with the generous outer margins of each page, please consider these blank pages as a canvas for your thoughts.

There are two editions of this work. The first is a paperback edition with a unique back cover-plate, produced with uncoated paper to encourage your notes. The second, a library edition produced as a case-bound volume with cloth binding and a shipping jacket to protect the cloth until delivery. There is no electronic edition of this work. This volume is intended to be held, marked, and kept as a living document with your observations.

The methods used to produce this book reflect the same principles argued within it: structure over convenience, permanence over speed, and truth over presentation.

Dedicated to the engineers who build infrastructure that lasts beyond applause.

About the Author

James L. Pulley III was born in Spartanburg County, South Carolina. By the sixth grade, he had lived in four cities and learned something that would shape everything that followed: that people can live next to the ocean and never see the beach. Proximity is not awareness. That observation became a career.

He graduated from Furman University in 1991 with a blended degree in computer science and business administration, and began his career at Microsoft, answering questions on the phone for Product Support Services. He became one of Microsoft's early electronic services voices on CompuServe. As an electronic services pioneer, this was the first of many times he would find himself explaining complex systems to people who needed help more than theory.

On April 1, 1996, he was hired as a field sales engineer at Mercury Interactive, the company that defined the performance testing industry. That date, and the discipline it introduced him to, became the origin coordinate of a thirty-year career in software performance engineering. April 1st was chosen for the launch of this book as the thirtieth anniversary of this hire date.

Over three decades, his work has spanned performance testing, capacity planning, application performance management, observability, and site reliability engineering. He has tested systems that governed a trillion dollars in mortgage assets, identified architectural flaws worth millions in monthly revenue, helped secure biometric entry records for every non-citizen entering the United States, and scaled a national digital wallet from ten thousand to one million concurrent users in four weeks. He has built teams, designed governance frameworks, and served as the person organizations called when the system was failing and the stakes were measured in minutes.

But the work he considers most important has always happened in the margins.

In high school, a neighbor's son lost a leg in a tragic accident during his junior year. James was asked to serve as his geometry tutor during recovery. That young man now builds custom prosthetic limbs for others who face the same challenge. The helped became the helper. It was the first time James saw the cycle that would define his professional life, but not the last.

On September 12, 2001, he began answering questions from peers in online technical forums. Not as strategy. Because it was the only useful thing available. Those answers, and the thousands that followed across the next two and a half decades, exceed the length of Homer's Odyssey. They are permanent. They are public. And they are the foundation on which every book, every framework, and every reputation in this body of work was built.

In 2008, he founded Journeyman Publishing — named for the guild tradition in which a journeyman is one who has completed an apprenticeship and earned the right to travel independently, practicing the craft wherever it is needed. He chose the title deliberately. While some of his peers see him as a master, he sees himself as a permanent learner on a journey that does not end.

In 2012, he co-founded PerfBytes, a podcast on software performance engineering inspired by the humor of CarTalk, the accessibility of Alton Brown's Good Eats, and the timing of British comedy. The show has been downloaded more than 177,000 times and continues today. It is proof that technical depth and human warmth are not opposites.

He serves on the board of the Advancing Technology Ventures Lab, a 501(c)(3) founded by Dennis Hayes to mentor early-stage ventures in upstate South Carolina. He has led, mentored, taught, and answered questions in every role he has held. In many cases, this was a task he was never asked to fill.

He often jokes that if he could extract a few cents from every success he has helped others achieve, he would never have to work again. When his father passed in 2006, he received handwritten letters from people around the world whose careers he had quietly shaped. Those letters confirmed what the work had always told him: that usefulness, practiced consistently, is the only authority that survives.

He is the author of *Navigating Performance Engineering*, *Software Performance Risk Management*, *The Fleetwood*, and the *Reputation Engineering* education series. His earlier work includes *Interviewing & Hiring Software Performance Test Professionals*, as well as dozens of articles and whitepapers across forums and conferences.

He lives in Spartanburg County, South Carolina, with his wife Rachel, in a home built around a log cabin from the late 1700s that they are expanding together, by hand. Rachel has been on this land, purchased by her parents, since she was six years old. At the same time James wandered sixteen cities across the country between 1976 and 2018 before arriving at the same three acres. Dorothy had it right in the *Wizard of Oz*, "There's no place like home..."

James has never believed that expertise is the point. The point is the person standing next to you who doesn't have it yet.

Throughout his career, he has helped companies understand why software fails under load, why performance results often lack defensible structure, and how risk emerges when measurement is treated as output rather than evidence.

His work has focused on helping engineering teams and business leaders ask better questions: What is the declared invariant? Can it repeat? What are the bounds of acceptable variance? What is the economic consequence if it fails?

Pulley's approach centers on restoring scientific discipline to performance engineering—emphasizing reproducibility, auditability, and the translation of technical results into business risk.

Risk is the probability that a declared performance
invariant
will fail under defined conditions,
multiplied by the economic consequence of that failure.

If it cannot be repeated, it cannot be priced.
If it cannot be priced, it cannot be governed.

www.ingramcontent.com/pod-product-compliance
Lightning Source LLC
Chambersburg PA
CBHW042047030726
47599CB00019B/2391